Measure of Central Tendency and Dispersion (With Solved Problems)

NIPA® GENX ELECTRONIC RESOURCES & SOLUTIONS P. LTD.
New Delhi-110 034

About the Authors

R. Sugunthakunthalambigai working as an Assistant Professor of Mathematics, Tamil Nadu Agricultural University, Agricultural Engineering College and Research Institute, Kumulur. She completed B.Sc. in D.G. Government Arts College for Women, Mayiladuthurai, M.Sc. in T.B.M.L. College of Arts and Science, Porayar, M.Phill (Algebra) in Government Arts College (Autonomous), Kumbakonam and Ph.D. (Fuzzy Mathematics) in T.B.M.L College of Arts and Science, Porayar, Bharathidasan University.

She has experienced professionally 17 years in Teaching, Research and Extension. She published more than 15 Research papers in National and International reputed journals in mathematics, 10 Nos. of Book chapters (ISBN). She received more than 10 awards including Academic Excellence Award, Professional Excellence Award, Dr.K.C Mehta Best Researcher Award, Best Young Scientist Award, Best Paper Award, Best Paper - Author Award in 8th National Conference on Agricultural Scientific Tamil, Nagapattinam organized by Agricultural Scientific Tamil Society, New Delhi, InRes Research Excellence Award, S Ramanujan Award for outstanding Scientist Award, Teaching Achievers Award and Excellence Teaching in Higher Education Award.

She is expert in Mathematics and allied subjects such as Sage Math, Block chain, Artificial Intelligence and machine learning, mathematical Prototypes, Pandas in Python, mat lab, Python, SPSS, R-Programme, Ethics research etc. She is expertise in mathematics to giving coaching to the Agricultural and allied subjects for ICAR and GATE examinations.

Dr.R.Brimapureeswaran working as an Assistant Professor of Food Process Engineering, Tamilnadu Dr. J. Jayalalithaa Fisheries University, College of Fish Nutrition and Food Technology, Chennai. He graduates a Ph.D. in Food Process Engineering from the Indian Institute of Food Processing Technology (NIFTEM) Thanjavur, Tamilnadu, M.Tech in Post Harvest Engineering from Indian Institute of Technology, Kharagpur, West Bengal and B.Tech in Agricultural Engineering from Agricultural Engineering College and Research Institute, Kumulur, Tamilnadu Agricultural University.

He has experienced professionally more than 10 years in Teaching, Research and Extension in the field of Food and Fish Process Engineering. He published more than 12 Research papers in National and International reputed journals and a Book chapters (ISBN). He received more than 10 awards including Rajiv Gandhi National Fellowship Award, Best Teacher Award, Research Peace Award, Best Researcher Award, Whole Grain 2015 Award, Best Young

Scientist Award, Best Faculty Award, NCOFTECH- II Award, ETAE Award and GATE qualifying Award. He has established various projects and centers related to food processing and has designed and developed several innovative food processing equipment.

He is expert in Food Engineering, Food Process Engineering, Food Storage Engineering, Food Packaging, Processing of Spices, Fish Process Engineering and Unit operations in Agricultural Process Engineering. He is an Associate editor of the Food Technology Research Journal, Egypt and also member of several esteemed Scientific bodies and holds additional academic positions with leadership abilities.

Measures of central tendency

Mean, median, mode - types of mean - Ungrouped or Raw Data formula with simple problems - Grouped data with simple problems (when class intervals not given, when class intervals given) - problems for the mean of the given data by step deviation method - Combined mean - Weighted mean - Merits, Demerits and uses of Arithmetic Mean - Definition of median - ungrouped - grouped (When the series is discrete and continuous) - Merits and Demerits of median-Definition of mode - simple problems for grouped and ungrouped-Merits and Demerits of mode-definition of Geometric Mean-problems for Discrete and continuous series-definition of Harmonic Mean with problems.

Measures of Dispersion

Types of dispersion - definition of Range and coefficient of range with problems - simple problems for Quartile Deviation - quartile deviation and its coefficient - problems for Mean Deviation from (i) mean (ii) mode (iii) median - Discrete and continuous series - coefficient of mean deviation - Definition of Standard Deviation - simple problems for S.D. and Co-efficient of variance-S.D. For discrete frequency distribution-variance - Co-efficient of variance-Characteristics of SD - Uses - Characteristic of CV.

The book contains solved problems of Measure of central tendency and measure of dispersion.

Measure of Central Tendency and Dispersion (With Solved Problems)

R. Sugunthakunthalambigai
Assistant Professor
Mathematics
Agricultural Engineering College and Research Institute
Tamil Nadu Agricultural University
Kumulur, Tamil Nadu

R. Brimapureeswaran
Assistant Professor
Food Process Engineering
College of Fish Nutrition and Food Technology
Tamil Nadu Dr. J. Jayalalithaa Fisheries University
Chennai, Tamil Nadu

NIPA® GENX ELECTRONIC RESOURCES & SOLUTIONS P. LTD.
New Delhi-110 034

NIPA® GENX ELECTRONIC RESOURCES & SOLUTIONS P. LTD.

101,103, Vikas Surya Plaza, CU Block
L.S.C. Market, Pitam Pura, New Delhi-110 034
Ph : +91-11-43860225, Mob.: +91 9717133558, 9540816132
E-mail: newindiapublishingagency@gmail.com
Website: www.nipaersources.com

Print ISBN: 978-93-58875-91-1
ebook ISBN: 978-93-58872-31-6

Composed and Designed by NIPA®.

Preface

Mathematics is the queen of the sciences and cornerstone of all aspects of Engineering. Hence, it is of prime importance that all budding engineers have a strong grasp of the basic principles of mathematics. This book explores the diverse facets of mathematics and its bound less applications and cater to needs of students such as Agriculture, Horticulture, Agricultural Engineering, Fisheries Engineering and common Engineering. This book has been written to meet the requirements of graduate students of all Indian universities. As far as possible mathematical complexity has been avoided and wherever necessary, the notations and terminologies have been clearly described with detailed explanation of mathematical steps involved. It designed to be user-friendly. Each chapter of this book has complete theory and large number of solved problems. It has been made to make the language simple and easily understandable which conforms to definitions and explanations. The matter has been gradually developed in logical and methodical manner involving as little as possible.

We are sure that it will be invaluable both to the students and the teacher alike. We welcome constructive comments and suggestions.

Authors

Contents

1

Measures of Central Tendency

1.1 Introduction

One of the important objectives of statistical analysis is to determine various measures. The first of such measures is average. In the words of Prof. R.A. Fisher", The inherent inability of a human **to grasp in its entirely a large body of numerical data compels us to seek relatively few constants that will adequately describe the data".**

- The mean value which is equal to the ratio of the sum of the number of a given set of values to the total number of values present in the set is known as average.
- Average has many applications in real-life.
- Suppose we need to find the average age of men or women in a village or average female height in India. We can calculate it by adding all the values and dividing it by the number of values we have added.
- Here's the average formula in math that evaluates the average $= \frac{\text{sum of elements}}{\text{Number of elements}}$

Every dataset has a property to tend towards a central point as the sample size tends towards infinity. This property of data is called as central tendency and the point towards data tends is called as Measures of central tendency. These measures can also be termed as the central point of data around which data is supposed to be scattered (or concentrated).

1.2. Measures of Central Tendency or Averages – some definition of differently by different statisticians

"Measures of central tendency (averages) are statistical constants which enable us to comprehend in a single effort the significance of the whole." - **Prof Bowley**

"A measure of Central tendency is a typical value around which other figures congregate." —**Simpson and Kafka**

"An average stands for the whole group of which it forms a part, yet represents the whole". —**A.E. Waugh**

"An average is a figure that represents the whole group." —**Clark**

"Statistics may be defined as the science of collection, presentation, analysis and interpretation of numerical data".- **Croxton and Cowden,**

1.3. The main measures of central tendency are

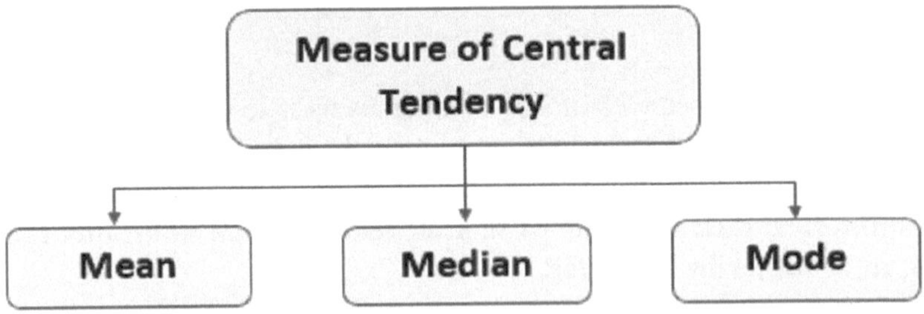

1.4. Specifications of a Good Average or Measure of Central Tendency:

a) It should be rigidly defined.

b) It should be simple to understand & easy to calculate.

c) It should be based upon all values of given data.

d) It should be capable of further mathematical treatment.

e) It should have sampling stability.

f) It should be not be unduly affected by extreme values.

1.5. Various Measure of Central Tendency

The following are the five Measure of Central Tendency are commonly used in practice.

1. Arithmetic mean: Regular meaning of "average"
   i) Simple Arithmetic mean
   ii) Combined Arithmetic mean
   iii) Weighted Arithmetic mean
2. Median: **Middle value**
3. Mode: **Most often**
4. Geometric mean
5. Harmonic mean

1.6. Arithmetic mean

Ungrouped or Raw Data

$$\bar{X} = \frac{\sum_{i=1}^{n} X_i}{n}$$

$\bar{X}$ = Arithmetic mean.

ΣX_i = sum of all the items of the variable.

n = **Number of observations.**

1.6.1. Calculate the arithmetic mean of the following data.10,20,30,40,50.

Solution

Mean = $\sum$x/n

= 10+20+30+440+50/5

= 150/5

= 30.

Mean= 30

1.6.2. Calculate the A.M for the following data.75,95,80,85,79,20,50.

Solution

Mean =$\sum$x/n

= 75+95+80+85+79+20+50/7

= 484/7

= 69.14.

Mean= 69.14.

1.6.3. Calculate the mean for the following data.22.5,30,35.5,42.7, 69.8,70.5,92.4.

Solution

Mean= $\sum$x/n

= 22.5+30+35.5+42.7+69.8+70.5+92.4/7

Mean= 51.91.

1.6.4. Calculate the mean for the following data 259,375,407,502,169,705.

Solution

Mean= $\sum$x/n

= 259+375+407+502+169+705/6

= 2417 /6

Mean= 402.8.

1.6.5. Calculate the mean for the following data.750.5,320.2,795.5, 362.4,525.3,272.2.

Solution

Mean= 750.5+320.2+795.5+362.4+525.3+272.2/6

= 3026.1/6

= 504.4

Mean= 504.4.

1.7. Grouped data : (i) when class intervals not given

$\bar{X} = \sum fX / \sum f$

$\sum fX$= sum of the products of the frequencies and the variable X.

$\sum f$= sum of the frequencies.

1.7.1. Calculate the arithmetic mean for the following data.

Value	10	20	30	40	50	60	70	80
Frequency	3	6	10	12	9	6	2	2

Solution

X	f	fX
10	3	30
20	6	120
30	10	300
40	12	480
50	9	450
60	6	360
70	2	140
80	2	160
	$\sum f$=50	$\sum fX$=2040

Mean = $\sum fX / \sum f$

= 2040/50

= 40.8

$\bar{X}$ (or) Mean = 40.8

1.7.2. Calculate the A.M for the following data.

Value	20	40	60	80	100
Frequency	3	4	7	9	2

Solution

X	f	fX
20	3	60
40	4	160
60	7	420
80	9	720
100	2	200
	∑f=25	∑fX=1560

Mean= ∑fX /∑f

= 1560/25

= 62.4

Mean= 62.4

1.7.3. Calculate the A.M for the following data.

Value	30	60	90	120	150
frequency	3	5	7	9	11

Solution

X	f	fX
30	3	90
60	5	300
90	7	840
120	9	1080
150	11	1650
	∑f=35	∑fX=3960

Mean=∑fX /∑f

= 3960/35

= 113.14

1.7.4. Calculate the A. M for the following data

Value	5	10	15	20	25	30
frequency	3	2	5	1	7	6

Solution

X	f	fX
5	3	15
10	2	20
15	5	75
20	1	20
25	7	175
30	6	180
	$\sum f=24$	$\sum fX=485$

Mean $= \sum fX / \sum f$

$= 485/24$

Mean $= 20.21$

1.7.5. Calculate the A.M for the following data

Value	54	69	75	82	95
frequency	7	8	4	2	5

Solution

X	f	fX
54	7	378
69	8	552
75	4	300
82	2	164
95	5	475
	$\sum f=26$	$\sum fX=1869$

Mean $= 1869/26$

$= 71.88$

Mean $= 71.88$

ii. when class intervals are given

$\bar{X} = \sum fm / \sum f$ (direct method)

$\sum fm$= sum of the products of the frequencies and the midpoints

$\sum f$= sum of the frequencies.

1.7.6. Calculate the A.M from the following data

MARKS: 0-10 10-20 20-30 30-40 40-50 50-60

NO.OF STUDENTS: 8 7 4 9 3

Solution

X(marks)	f(frequency)	M(mid points)	fm
0-10	8	5	40
10-20	7	15	105
20-30	4	25	100
30-40	9	35	315
40-50	3	45	135
50-60	6	55	330
	∑fx=37		∑fm= 1025

A.M= ∑fm/∑f

= 1025/37

A.M= 27.7

1.7.7. Calculate the A.M from the following data

X	10-19	20-29	30-39	40-49	50-59	60-69
f(frequency)	12	19	31	27	16	8

Solution

x	f(frequency)	m(mid points)	fm
10-19	12	14.5	174
20-29	19	24.5	465.5
30-39	31	34.5	1069.5
40-49	27	44.5	1201.5
50-59	16	54.5	872
60-69	8	64.5	516
	∑fx=113		∑fm=4298.5

A.M= ∑fm/∑fx

= 4298.5/113

A.M= 38.03

1.7.8. Calculate the A.M from the following data

x (doses of fertilizer)	0-6	6-12	12-18	18-24	24-30	30-36
f(yield)	12	24	36	38	33	6

Solution

x(doses of fertilizer)	f(yield)	m(mid point)	fm
0-6	12	3	36
6-12	24	9	216
12-18	36	15	540
18-24	38	21	798
24-30	33	27	891
30-36	6	33	198
	fx=149		fm=2643

A.M =∑fm/∑fx

=2643/149

A.M =17.74

1.7.9. Calculate the A.M from the following data

x(nitrogen input)	0-10	10-20	20-30	30-40	40-50	50-60	60-70	70-80
F(output)	4	12	20	18	15	8	2	1

Solution

x(nitrogen)input	f(yield) output	m(midpoint)	fm
0-10	4	5	20
10-20	12	15	180
20-30	20	25	500
30-40	18	35	630
4050	15	45	675
50-60	8	55	440
60-70	2	65	130
70-80	1	75	75
	∑fx=80		∑fm=2575

A.M= ∑fm/∑fx

= 2575/80

A.M= 32.18

1.7.10. Calculate the A.M from the following data

X(mop in kg)	0-10	10-20	20-30	30-40	40-50
F(yield)	7	12	24	10	7

Solution

X(mop in kg)	F(yield)	M(midpoint)	fm
0-10	7	5	35
10-20	12	15	180
20-30	24	25	600
30-40	10	35	350
40-50	7	45	315
	∑fx=60		∑fm=1480

A.M= ∑fm/∑fx

= 1480/60

A.M= 24.

1.7.11. Calculate the A.M from the following data

x	0-5	5-10	10-15	15-20	20-25
f	10	20	30	40	50

Solution

x	f(frequency)	m(midpoint)	fm
0-5	10	2.5	25
5-10	20	7.5	150
10-15	30	12.5	375
15-20	40	17.5	700
20-25	50	22.5	1125
	∑fx=150		∑fm=2375

A.M= ∑fm/∑fx

= 2375/150

A.M= 15.8

1.7.12. Calculate the A.M from the following data

x(class)	0-10	10-20	20-30	30-40	40-50
F(frequency)	25	30	35	20	15

Solution

x(class)	f(frequency)	m(midpoint)	fm
0-10	25	5	125
10-20	30	15	450
20-30	35	25	875
30-40	20	35	700
50-60	15	45	675
	∑fx=125		∑fm=2825

A.M= ∑fx/∑fm

= 2825/125

A.M= 22.6

1.7.13. Calculate the A.M from the following data

x(class)	10-20	20-30	30-40	40-50	50-60
f(frequency)	5	9	13	21	20

Solution

x(class)	**F(frequency)**	**m(midpoint)**	**fm**
10-20	5	15	75
20-30	9	25	225
30-40	13	35	455
40-50	21	45	945
50-60	20	55	1100
	∑fx=68		∑fm=2800

A.M= ∑fx/∑fm

= 2800/68

A.M= 41.17

1.7.14. Calculate the A.M from the following data

x(class)	10-20	20-30	30-40	40-50
f(frequency)	21	13	9	15

Solution

x(class)	**f (frequency)**	**m(midpoint)**	**fm**
10-20	21	15	315
20-30	13	25	325
30-40	9	35	315
40-50	15	45	675
	∑fx=58		∑fm=1630

A.M= ∑fx/∑fm

= 1630/58

A.M = 28.1

1.7.15. Calculate the A.M from the following data

X(class)	0-10	10-20	20-30	30-40	40-50
f(frequency)	5	8	15	13	9

Solution

X(class)	f(frequency)	m(midpoint)	fm
0-10	5	5	25
10-20	8	15	120
20-30	15	25	375
30-40	13	35	455
4050	9	45	405
	∑fx=50		∑fm=1380

A.M= ∑fx/∑fm

= 1380/50

A.M= 27.6

1.8. Step deviation method (another method)

$$\text{Mean} = A + h.\frac{\sum fi\,di}{\sum fi}$$

A = Assumed mean

$$d = \frac{X - A}{h}$$

h = class interval

Note: when class interval is not taken

$$\overline{X} = A + \frac{\sum fd}{n}$$

where

d = X-A

1.8.1. Calculate the mean of the given data by step deviation method.

class	32.5	37.5	42.5	47.5	52.5	57.5	62.5	67.5	72.5	77.5
freuency	10	8	18	3	4	19	18	3	5	2

Solution

A = Assumed mean = 57.5, h = 5

xi	fi	$di\frac{xi - A}{n}$	fi di	$\lvert xi - M \rvert$	$fi\lvert xi - M \rvert$
32.5	10	-5	-50	19.5	195
37.5	8	-4	-32	14.5	116
42.5	18	-3	-54	9.5	171
47.8	3	-2	-6	4.5	13.5
52.5	4	-1	-4	0.5	2.0
57.5	19	0	0	5.5	104.5
62.5	18	1	18	10.5	189
67.5	3	2	6	15.5	46.5
72.5	5	3	15	20.5	102.5
77.5	2	4	8	25.5	51
	90		-99		991

$$\text{Mean} = A + h.\frac{\Sigma\,fidi}{\Sigma\,fi}$$

$$= 57.5 - (5)\frac{99}{90} = 52$$

$\therefore$ Mean= 52

1.9. COMBINED MEAN

$$\bar{z} = \frac{n_X \bar{X} + n_y \bar{y}}{n_X + n_y}$$

nx,ny = number of people

$\bar{x}, \bar{y}$ = means

1.9.1. Calculate the mean of the given data

group first the number of people -(n_x) = 35

mean - $\bar{x} = 75$

group second the number of people - (n_y) = 55

mean - $\bar{y} = 95$

Solution

$$\bar{z} = \frac{n_X \bar{X} + n_y \bar{y}}{n_X + n_y}$$

combined mean = (35*75+55*95) / (35+55) = 87.22

1.10. WEIGHTED MEAN

The weighted mean M_w of a set of N values (x_1, x_2,..., x_N)is computed according to the following formula

$$Mw(x1, x2, ..., xN) = \frac{\sum_{i=1}^{i=N} w_i \cdot x_i}{\sum_{i=1}^{i=N} wi}; w_i \geq 0.$$

where

W_1, W_2,...W_N are non-negative coefficients, called "weights", that are ascribed to the corresponding values x_1, x_2,...,x_N.

Only the relative values of the weights W_i matter in determining the value of the weighted mean. In other words, if you multiply each weight W_i by a positive value K and recalculate the weighted mean with new weights (K· W_1,..., K·W_N), then the value M_w remains the same.

The greater the weight W_i in respect to other weights, the greater the contribution of the corresponding x_i datum into the resultant value M_w of the weighted mean.

The weighted mean is used, for example, to aggregate a set of scores (e.g. examination scores on different subjects) to a single resultant score. In this case, the magnitude of the weights W_i reflects the importance of their respective subjects.

(For this reason, the weighted mean is sometimes named the aggregate mean).

1.10.1. Calculate weighted mean: we had two school classes, one with 20 students, and one with 30 students. The grades in each class on a particular test were:

Morning class = 62, 67, 71, 74, 76, 77, 78, 79, 79, 80, 80, 81, 81, 82, 83, 84, 86, 89, 93, 98

Afternoon class = 81, 82, 83, 84, 85, 86, 87, 87, 88, 88, 89, 89, 89, 90, 90, 90, 90, 91, 91, 91, 92, 92, 93, 93, 94, 95, 96, 97, 98, 99

Solution

The straight average for the morning class is 80% and the straight average of the afternoon class is 90%. If we were to find a straight average of 80% and 90%, we would get 85% for the mean of the two class averages. However, this is not the average of all the students' grades. To find that, you would need to total all the grades and divide by the total number of students:

$\bar{x} = 4300 / 50 = 86\%$

Or, you could find the weighted average of the two class means already calculated, using the number of students in each class as the weighting factor:

$\bar{x} = (20) * 80\% + (30) * 90\% \ / \ 20 + 30$

$= 1600 + 2700 / 50$

$= 4300 / 50 = 86\%$

Note that if we no longer had the individual students' grades, but only had the class averages and the number of students in each class, we could still find the mean of all the students grades, in this way, by finding the weighted mean of the two class averages.

1.11. Merits, Demerits and uses of Arithmetic Mean

1.11.1. Merits

- It can be easily calculated,
- calculation is based on all the observations,
- easy to understand,
- rightly defined by the mathematical formula,
- least affected by sampling fluctuations,
- the best measure to compare two or more series,
- the average obtained by calculations and it does not depend upon any position.

1.11.2. Demerits

- It may not be represented in actual data so it is theoretical.
- The extreme values have greater effect on mean.
- It cannot be calculated if all the values are not known.
- It cannot be determined for the qualitative data such as good, bad, beauty, honesty etc.
- Mean may lead to fallacious conditions in the absence of original observations.

1.11.3. Uses of Arithmetic Mean

- A common man uses mean for calculating average marks obtained by a student.
- It is extremely used in practical statistics.
- Estimates are always obtained by mean.

- Businessman uses it to find out the operation cost, profit per unit of article.

2. Median

The median is the middle most items that divides the distribution into two equal parts when the items are arranged in ascending order.

When the data is ungrouped

Arrange the n values of the variable in ascending (or) descending order of magnitudes.

Case I :- When n is odd. In this case $\frac{n+1}{2}$ th value is the median.

i.e Median = $\frac{n+1}{2}$ th term.

2.1. Calculate the median from following data.

X= 5,19,6,16,15.

Solution

X= 5,6,15,16,19.

Here n = 5.('n' is odd.)

Median = $\frac{n+1}{2}$ th term.

= $\frac{5+1}{2}$ th term.

=3rd term =6

Median=6.

2.2. If the number of panicle per rice plant is 18,16,20,12,15.

Solution

X=12,15,16,18,20.

Here n = 5.

Median = $\frac{n+1}{2}$ th term.

= $\frac{5+1}{2}$ th term.

=3rd term =16.

Median =16 panicle per plant.

2.3. Calculate the median from the following series.

X=25,20,30,35,15,40,45.

Solution

X=15,20,25,30,35,40,45.

Here n=7.

Median $\frac{n+1}{2}$ th term.

$=\frac{7+1}{2}$ th term.

$=4^{th}$ term=15.

Median=15.

2.4. The sorghum produced cobs per 5m² is 30,27,25,23,21,28,31.

Solution

X=21,23,25,27,28,30,31.

Here n =7.

Median= $\frac{n+1}{2}$ th term.

$=\frac{7+1}{2}$ th term.

$=4^{th}$ term=27.

Median=27 cobs per 5m².

2.5. Calculate the median from the following series.

X = 89,85,90,95,93,72,65,78,77.

Solution

X=65,72,77,78,85,89,90,93,95.

Here n=9.

Median = $\frac{n+1}{2}$ th term.

$=\frac{9+1}{2}$ th term.

$= 5^{th}$ term=85.

Median=85.

2.6. In gingely the oil content was 50%, 56%, 35%, 32%, 28%, 45%, 48%.Calculate the median.

Solution

X=28,32,35,45,48,50,56.

Here n=7.

Median= $\frac{n+1}{2}$ th term.

$=\frac{7+1}{2}$ th term.

=4th term =45.

Median =45%.

Case II:- When n is even, in this case there are two middle terms $\frac{n}{2}$ th and $\frac{n}{2}$ +1 th

The median is the average of these two terms Median $=\frac{\frac{n}{2}+\left(\frac{n}{2}+1\right)}{2}$

2.7. Calculate the median from following from data.

X=10, 3,7,5,6,19,12,13.

Solution

X=3, 5, 6, 7, 10, 12, 13, 19. (Arrange the series in ascending order)

Here N=8.('n' is even)

Median= $\frac{n}{2}$ & $\frac{n}{2}$ +1th term.

Median=8/2 & $\frac{8}{2}$+1th term

=4 & 5th term

4th item= 7 & 5th item=10

Median=7+10/2

=17/2

Median=8.5

2.8. If the height of the rice crops are 35,40,42,45,37,46cm.

Solution

X=35,37,40,42,45,46.

Here n=6.

Median= $\frac{n}{2}$ & $\frac{n}{2}$ +1th term.

$=\frac{6}{2}$ & $\frac{6}{2}$+1th term.

=3rd & 4th term

3rd term=40 & 4th term =42.

Median=40+42/2

=82/2

Median=41cm.

2.9. Calculate the median from following data.

X=48,45,40,64,60,55.

Solution

X=40,45,48,55,60,64.

Here n =6

Median= $\frac{n}{2}$ & $\frac{n}{2}$ +1th term.

=6/2 & $\frac{6}{2}$ +1th term.

=3rd & 4th term.

3rd term=48 & 4th term=55.

Median=48+55/2

=103/2.

Median =51.5

2.10. If the sunhemp plant per m² is 15,11,20,16,18,10,14,22,24,21.

Solution

X=10,11,14,15,16,18,20,21,22,24.

Here n=10.

Median = $\frac{n}{2}$ & $\frac{n}{2}$ +1th term.

=10/2 & 10/2+1th term.

=5th & 6th term.

5th term=16 & 6th term=18.

Median=16+18/2

=34/2

Median=17 Plants per m^2.

2.11. Find out the following for the data

1, 2, 3,4,5,6,

Solution

Here 'n' is 6

Median = $\frac{n}{2}$ & $\frac{n}{2}$ +1th term.

=6/2 & 6 th term

=3 & 4 th term

So, 3+4/2

Median = 3.5

2.1.0. When the data is grouped

Case (i) When the series is discrete.

In this case the values of the variable are arranged in ascending (or) descending order of magnitudes.

Median = size of $\frac{(n+1)}{2}$ th item where n= $\sum f$= total frequencies

2.1.1. Calculate median for the following data

No.of students	6	4	16	7	8	2
Marks	20	9	25	50	40	80

Solution

Arranging the marks in ascending order and preparing the table.

Marks	Frequency	Cumulative frequency(c.f)
9	4	4
20	6	10
25	16	26
40	8	34
50	7	41
80	2	43
	n=$\sum$f=43	

Here n = 43, Median $= \frac{(n+1)}{2} =$ th value

$= \frac{(43+1)}{2}$ the value = 22nd value

The above table shows that all items from 11 to 26 have their values 25. Since 22nd item lies in this interval, therefore the value is 25. Hence Median = 25 marks.

2.1.2. Calculate the median for the following data

Marks	5	7	9	12	8
No of students	4	3	9	7	5

Solution

X	f	c.f
5	4	4
7	3	7
8	5	12
9	9	21
12	7	28
	Σf= n = 28	

Here n = 28, Median = $\frac{(n+1)}{2}$ th value

= $\frac{(28+1)}{2}$ the value = 14.5 th value

The above table shows that all items from 13 to 21 have their values 9. Since14.5 th item lies in this interval, therefore the value is 9. Hence Median = 9 marks.

2.1.3. Calculate the median for the following data

x	8	9	11	12	10
f	3	6	5	4	3

Solution

x	f	cf
8	3	3
9	6	9
10	3	13
11	5	18
12	4	21

Here Σf= n = 21, Median = $\frac{(n+1)}{2}$ th value

= $\frac{(21+1)}{2}$ the value = 11th value

Median = 10

2.1.4. Calculate the median for the following data

x	7	8	6	5	4
f	3	6	8	5	3

Solution

x	f	cf
4	3	3
5	5	8
6	8	16
7	3	19
8	6	25
	Σf= n = 25	

Here Σf= n = 25, Median = $\frac{(n+1)}{2}$ th value

$= \frac{(25+1)}{2}$ the value = 13th value

Median = 6

2.1.5. Calculate the median for the following data

x	8	6	5	9	10
f	4	7	3	11	8

Solution

x	f	cf
5	3	3
6	7	10
8	4	14
9	11	25
10	8	Σf= n =33

Here Σf= n = 33, Median = $\frac{(n+1)}{2}$ th value

$= \frac{(33+1)}{2}$ the value = 17th value

Median = 6

2.2.0 Case (ii) When the series is continuous

In this case the data is given in the form of a frequency table with class-interval

$$\text{Median} = L + \frac{\frac{\sum f}{2} - c.f}{f} xc$$

where

L = lower limit of the class in which the median lies

$\sum f$ = total number of frequencies

f = frequency of the class in which the median lies.

c.f = cumulative frequency of the class preceding the median class

c = width of the class interval of the class in which the median lies.

2.2.1. Calculate the median for the following data

Class	10-19	20-29	30-39	40-49	50-59	60-69
frequency	12	19	31	27	16	8

Solution

Class	**f**	**c.f**
10-19	12	12
20-29	19	31
30-39	31	62
40-49	27	89
50-59	16	105
60-69	8	113
	$\sum f = n = 113$	

L = lower limit of the class in which the median lies = (29.5)

$\sum f$ = total number of frequencies = 113

f = frequency of the class in which the median lies.=31

c.f = cumulative frequency of the class preceding the median class = 31

c = width of the class interval of the class in which the median lies = 10

Median= size of $\sum f / 2$ th item=113/2=56.5th item,

Median lies in 30-39

But the real limit of this class is 29.5-39.5

$$\text{Median} = L + \frac{\frac{\sum f}{2} - c.f}{f} \times c.$$

$$= 29.5 + \frac{56.5 - 31}{31} \times 10$$

=29.5+(.822 X10)

=37.73

Median is 37.73

2.2.2. Calculate the median for the following data

Class	10-20	20-30	30-40	40-50	50-60	60-70	70-80	80-90
frequency	5	9	13	21	20	15	8	3

Solution

class	**f**	**cf**
10-20	5	5
20-30	9	14
30-40	13	27
40-50	21	48
50-60	20	68
60-70	15	83
70-80	8	91
80-90	3	94
	$\sum f = 94$	

$$\text{Median} = L + \frac{\frac{\sum f}{2} - cf}{f} \times c$$

$\sum f/2 = 94/2 = 47^{th}$ term,

L=40, c.f = 27, c =10,f= 21

$$\text{Meddian} = L + \frac{\frac{94}{2} - 27}{21} \times 10$$

$$= 40 + \frac{20}{21} \times 10$$

Median =49.5

2.2.3.Calculate the median for the following data

Value	Less than 100	100-200	200-300	300-400	400 and above
Frequency	50	90	158	68	134

Solution

class	f	cf
Less than 100	50	50
100-200	90	140
200-300	158	298
300-400	68	366
400 and above	134	500

Median= size of $\sum f/2$ th item=500/2=250th item,

Median lies in the class 200 - 300

$$\text{Median} = L + \frac{\frac{\sum f}{2} - c.f}{f} \times c$$

$\sum f/2$ =500/2 =250th term,

L=200, c.f= 140, c =100,f= 158

$$\text{Median} = 200 + \frac{250 - 140}{158} \times 100$$

=200+69.62

Median=269.62

2.2.4. The following table gives you the distribution of marks secured by some students in the examination

Marks	0-20	21-30	31-40	41-50	51-60	61-70	71-80
No. of students	42	38	120	84	48	36	31

Solution

class	f	c.f
0-20	42	42
21-30	38	80
31-40	120	200
41-50	84	284
51-60	48	332
61-70	36	368
71-80	31	399
	$\sum f$= n =399	

Median= size of $\sum f/2$ th item=399/2=199.5th item,

Median lies in the class 31-40

But the real limit of this class is 30.5-40.5

$L = 30.5$

$\sum f = 399$

$f = 120$

$c.f = 80$

$c = 10$

$$\text{Median} = L + \frac{\frac{\sum f}{2} - c.f}{f} \times c.$$

$$= 30.5 + \frac{199.5 - 80}{120} \times 10$$

=30.5+9.96

=40.46

Median is 40.46

2.3.0 Merits and Demerits of median

2.3.1 Merits

- Easily understood
- Not affected by extreme values
- Located graphically
- Best measure for qualitative data
- Easily located even if the class-intervals in the series are unequal
- It can be determined even by inspection in many cases

2.3.2 Demerits

- It is not subject to algebraic treatments
- It cannot represent the irregular distribution series.

3.0. Mode

The mode is the most frequent data value. Mode is the value of the variable which is predominant in the given data series. Sometimes there may be no single mode if no one value appears more than any other. There may also be two modes (bimodal), three modes (trimodal), or more than three modes (multi-modal).

For the raw data mode is that value which repeat most number of times. Thus if more than one observation repeated same number of times and is the most then we have more than one mode. If no observation repeated more than one time then the mode is ill-defined. For the discrete frequency distribution, mode is the value corresponding to maximum frequency.

Empirical formula: For symmetric distribution Mean, Median and Mode coincide. If the distribution is moderately asymmetrical the Mean, Median and Mode satisfy the following relationship

Mean- Mode =3(Mean- Median) Or Mode=3Median-2Mean

3.1: Find the mode from the following data

4,2,1,3,2,2,4,1,10,11,12

Solution: Number 2 is repeated in maximum level. So the mode is 2

3.2: Find the mode from the following data

4,4,4,2,1,3,2,2,4,1,10,11,12

Solution: Number 4 is repeated in maximum level. so the mode is 4

3.3: Find the mode from the following data.

40,44,57,68,78,67

Solution: In this case any number is not repeated. So no mode.

3.4: Find the mode from the following data.

45,55,50,45,40,55

Solution: here both numbers 45 & 55 occurs 2 times. So the mode 45 & 55

3.5: Find the mode from the following data.

22,42,34,14,45,56,33,12

Solution: any number is not come repeat. So no mode

3.6.The following is the number of problems that Ms.Matty assigned for homework on 10 different days. What is the mode ?

8,11,14,9,6,9,18,9,10,15

Solution: Ordering the data from least to greatest, we get

6, 8, 9, 9, 9, 10,11,14, 15, 18

Answer: The mode is 9

3.7. A marathon race was completed by 5 participants. What is the mode of these times given in hours?

2.7 hr,8.3 hr,3.5 hr, 5.1 hr 4.9 hr

Solution: Ordering the data from least to greatest, we get

2.7, 3.5, 4.9, 5.1, 8.3

Answer: no mode. Because any number is not repeated

3.8.On a cold winter day in January, the temperature for 9 North American cities is recorded in Fahrenheit. What is the mode of these temperatures?

1°,0°,0°,-1°,-5°,2°,6°,0°

Solution: The mode is 0°.it is recorded more times.

3.9. The number of points scored in a series of football games is listed below. Which score occurred most often?

7,13, 18, 24, 9, 3,18

Solution: Ordering the scores from least to greatest, we get

3, 7, 9, 13, 18, 18, 24

Answer

The score which occurs most often is 18.So the mode is 18

3.10. In a crash test, 11 cars were tested to determine what impact speed was required to obtain minimal bumper damage. Find the mode of the speeds given in miles per hour below?

24,15,18,20,18,24,26,18,26,24

Solution: Ordering the data from least to greatest, we get

15,18,18,18,20,22,24,24,24,26,26

Answer

Since both 18 and 24 occur three times, the modes are 18 and 24 miles per hour.

3.1.0 Calculate the mode for grouped data

For grouped frequency distributions, the modal class is the class with the largest frequency. After identifying modal class mode is evaluated by using formula.

Step1. Calculation of mean

Step2. Find out the modal class. The modal class is the class against the maximum frequency.

Step 3. Mode $= L + \frac{\Delta 1}{\Delta 1 \Delta 2} \times c$

Where L= lower limit of the modal class

$\Delta_1 = f_1 - f_0$

$\Delta_2 = f_1 - f_2$

f_1 = the frequency of the model group

f_0 = the frequency of the class preceding the modal class

f_2 = the frequency of the class succeeding the modal class

c = class interval

3.1.1. Calculate mode for given data

Urea	5-10	10-15	15-20	20-25	25-30	30-35
Output	7	10	13	21	20	26

Solution

Urea in kg (f)	Output(f)	cf
5-10	7	7
10-15	10	17
15-20	13(f_0)	30
20-25	21(f_1)	51
25-30	20(f_2)	71
30-35	26	97

$\sum f/2 = 97/2 = 48.5$ item.

Model group is 20-25

$$\text{Mode} = L + \frac{\Delta 1}{\Delta 1 \Delta 2} \times c$$

Where L= lower limit of the modal class= 20

f_1 = the frequency of the model group

f_0 = the frequency of the class preceding the modal class

f_2 = the frequency of the class succeeding the modal class

$\Delta_1 = f_1 - f_0 = 21-13=8$

$\Delta_2 = f_1 - f_2 = 21-20=1$

c = class interval =5

$$\text{Mode} = 20 + \frac{8}{8+1} \times 5$$

= 20+4.4

Mode = 24.4

3.1.2. Find out the mode for the following data

Family(x)	10-20	20-30	30-40	40-50	50-60
Members (f)	5	9	13	21	20

Solution

x	f	cf
10-20	5	5
20-30	9	14
30-40	13(f_0)	27
40-50	21(f_1)	48
50-60	20(f_2)	68
	$\sum f$=68	

$\sum f/2$= 68/2 = 34 item.

Model group is 40-50

$\Delta_1 = f_1 - f_0 = 21-13=8$

$\Delta_2 = f_1 - f_2 = 21-20=1$

c = class interval =5

$$\text{Mode} = L + \frac{\Delta 1}{\Delta 1 + \Delta 2} \times c$$

$$\text{Mode} = 40 + \frac{8}{8+1} \times 10$$

$$= 40 + \frac{8}{8+1} \times 10$$

= 40+8.8

= 48.8

3.1.3. Calculate the mode for the following table

Nitrogen to crop	2-10	10-18	18-26	26- 34	34-42	42-50
yields	4	7	12	20	18	11

Solution

Nit to crop	Yield	Cf
2-10	4	4
10-18	7	11
18-26	12(f_0)	23
26-34	20(f_1)	43
34-42	18(f_2)	61
42-50	11	72
	$\sum f$= 72	

$\sum f/2$ = 72/2 = 36 item.

Model group is26-34

$\Delta_1 = f_1 - f_0 = 20\text{-}12 = 8$

$\Delta_2 = f_1 - f_2 = 20\text{-}18 = 2$

c = class interval =8

$$\text{Mode} = L + \frac{\Delta 1}{\Delta 1 + \Delta 2} \times c$$

$$= 26 + \frac{8}{8+2} \times 8$$

$= 26 + 6.4$

$= 32.4$

3.1.4. Calculate the mode for following table

Organic manure X	5-10	10-15	15-20	20-25	25-30	30-35	35-40	40-45
Yield f	2	6	9	12	15	20	18	24

Solution

Organic manure (x)	**Yield(f)**	**c.f**
5-10	2	2
10-15	6	8
15-20	9	17
20-25	12	29
25-30	15(f_0)	44
30-35	20(f_1)	62
35-40	18(f_2)	82
40-45	24	$\sum f$=106

$\sum f/2 = 106/2 = 53$

Model group is 30-35

$\Delta_1 = f_1 - f_0 = 20\text{-}15 = 5$

$\Delta_2 = f_1 - f_2 = 20\text{-}18 = 2$

c = class interval =5

$$\text{Mode} = L \frac{\Delta 1}{\Delta 1 + \Delta 2} \times c$$

$$= 30 + \frac{5}{5+2} \times 5$$

$= 30 + 3.5$

Mode = 33.5

3.1.5. Compute the mode for following table

Irrigation	0-6	6-12	12-18	18-24	24-30
Hours	3	6	10	16	13

Solution

(Irrigation) x	**(Hours) f**	**cf**
0-6	3	3
6-12	6	9
12-18	10(f_0)	19
18-24	16(f_1)	32
24-30	13(f_2)	$\sum f = 48$

$\sum f/2 = 48/2 = 24$

Model group is 18-24

$\Delta_1 = f_1 - f_0 = 16\text{-}10 = 6$

$\Delta_2 = f_1 - f_2 = 16\text{-}13 = 3$

c = class interval = 6

$$\text{Mode} = L + \frac{\Delta 1}{\Delta 1 + \Delta 2} \times c$$

$$= 18 + \frac{6}{6+3} \times 6$$

$= 18 + 4$

$= 22$

3.1.6. Calculate the mode for the following data

Marks	1-10	10-19	19-28	28-37	37-46
No. of student	7	14	19	25	20

Solution

Marks (x)	**No. of students(f)**	**Cf**
1-10	7	7
10-19	14	21
19-28	19(f_0)	40
28-37	25(f_1)	65
37-46	20(f_2)	85

$\sum f = 85$

$\sum f/2 = 85/2 = 42.5$

Model group is 28-37

$\Delta_1 = f_1 - f_0 = 25\text{-}19 = 6$

$\Delta_2 = f_1 - f_2 = 25\text{-}20 = 5$

c = class interval = 9

$$\text{Mode} = L + \frac{\Delta 1}{\Delta 1 + \Delta 2} \times c$$

$$= 28 + \frac{6}{6+5} \times 9$$

= 28 + 4.9

Mode = 32.9

3.1.7. Find out the mode for the following data

No. of wages	0-4	4-8	8-12	12-16	16-20	20-24	24-28
Money in (Rs.)	5	10	14	18	23	17	27

Solution

x	f	cf
0-4	5	5
4-8	10	15
8-12	14	29
12-16	18(f_0)	47
16-20	23(f_1)	64
20-24	17(f_2)	87
24-28	27	$\sum f$ = 114

$\sum f/2 = 114/2 = 57$

Model group is 16-20

$\Delta_1 = f_1 - f_0 = 23\text{-}18 = 5$

$\Delta_2 = f_1 - f_2 = 23\text{-}17 = 6$

C = class interval = 4

$$\text{Mode} = L + \frac{\Delta 1}{\Delta 1 + \Delta 2} \times c$$

$$= 16 + \frac{5}{5+6} \times 4$$

= 16 + 1.8

Mode = 17.8

3.2.0 Merits and Demerits of mode

Merits

- Easily understood
- Can be located in some cases by inspection
- Capable of being ascertained graphically
- Not affected by extreme values.
- Represents the most frequent value
- The arrangement of data is not necessary if the items are a few

Demerits

- There are different formulae for its calculations which ordinarily given different answers.
- Mode is determinate. Some series have two (or) more than two modes.

4.0 Geometric Mean

The geometric mean is defined as the nth root of the product if n items of the series. The geometric mean of the series of $x_1, x_2, x_3 . . . x_n$ is $\sqrt[n]{x_1, x_2, ...x_n}$

Another method

The geometric mean is defined as the anti log of the arithmetic average of the logs of all the items of a series.

For raw data

$$G.M = \text{Anti}\log\left[\frac{\log x_1 + \log x_2 + ... + \log x_n}{N}\right]$$

$$G.M = \text{Anti}\log\left[\frac{\sum_{i=1}^{n} \log x}{N}\right]$$

4.1. What is the geometric mean of 4, 9, 9, and 2?

Solution

Just multiply the four numbers and take the 4th root:

$$\sqrt[4]{4 \times 9 \times 9 \times 2} = \sqrt[4]{648} = 5.045$$

4.2. What is the geometric mean of 2, 8 and 4?

Solution

Multiply those numbers together. Then take the third root (cube root) because there are 3 numbers.

$\sqrt[3]{2\times8\times4}=4$

4.3. Calculate the geometric mean for the following data 215.5, 6.5, 169, 11, 112.5, 14.2, 75, 35.5,

x	Log x
215.5	2.3334
6.5	0.8129
169	2.2279
11	1.0414
112.5	2.0511
14.2	1.1522
75	1.8750
35.5	1.5502
	$\sum$ Log x = 13.045

Solution

Geometric mean = antilog of $\sum$ Log x/N

Here N = number of observations = 8

$\sum$ Log x = 13.045

G.M = antilog of 13.045/8 = antilog of (1.630)

$\therefore$ G.M = 42.66

4.4. Calculate the geometric mean for the following data

50, 72, 54, 82, 93

X	Log x
50	1.6990
72	1.8573
54	1.7324
82	1.9138
93	1.9685
	$\sum$ Log x =9.1710

Solution

Geometric mean = antilog of $\sum$ Log x/N

Here N = number of observations = 5

$\sum$ Log x = 9.1710

G.M = antilog of 9.1710/5

= antilog of 1.8342

= 68. ∴ G.M = 68.26

4.5. Calculate the geometric mean for the following data 6, 9, 5, 7, 3

Solution

X	Log x
6	0.778
9	0.954
5	0.6990
7	0.8450
3	0.4771
	$\sum$ Log x = 3.7531

Geometric mean = antilog of $\sum$ Log x/N

Here N = number of observations = 5

$\sum$ Log x = 3.7531

G.M = antilog of 3.7531/5

= antilog of 0.75062

= 5.631

∴ G.M = 5.631

4.6. The Geometric mean of 3,7,8,9,14,18 and 20 is $\sqrt[7]{3x7x8x9x14x18x20}$

4.1.0 For Discrete series

$$G.M = \text{Anti}\log\frac{\sum f \log x}{\sum f}$$

4.1.1. Calculate the geometric mean for the following data

X	10	20	30	40	50	60
f	12	15	25	10	6	2

Solution

x	f	Log x	f log x
10	12	1.000	12
20	15	1.3010	19.515
30	25	1.4771	36.9275
40	10	1.6020	16.020
50	6	1.6990	10.194
60	2	1.7282	3.5564
	∑f =N= 70		∑f Log x =98.2129

G.M = antilog of ∑f Log x/N

= antilog of (98.2129/70)

= 1.4030

G.M = antilog of 1.4030

= 25.29

G.M = 25.29

4.1.2. The following table gives the weight of 31 persons in a sample survey. Calculate G.M.

Weight(lbs)	130	135	140	145	146
No. of persons	6	6	7	6	6

Solution

Weight	Frequency	Log x	F log x
130	6	2.1139	6.3417
135	6	2.1303	8.5212
140	7	2.1461	12.8766
145	6	2.1614	12.9684
146	6	2.1644	6.4932
	∑f = N=31		∑f Log x =47.2011

G.M = antilog of ∑f Log x/N

= antilog of (47.2011/31)

= antilog of 1.522

G.M = 33.26

4.1.3. Find out geometric mean

Income	14000	16000	17000	13000	15000
No. of families	11	22	14	15	13

Solution

Income x	F	Log x	f log x
14000	11	4.1461	45.60
16000	22	4.2041	92.49
17000	14	4.2304	59.22
13000	15	4.1139	61.70
15000	13	4.1760	54.28
	75		∑f Log x =313.29

G.M = antilog of ∑f Log x/N

= antilog of (313.29/75)

= antilog of (4.1772)

G.M = 15038.34

G.M = 15038.34

4.2.0 For continuous series

$$G.M = \text{Antilog} \frac{\sum f \log m}{\sum f} \text{ where m = mid-point}$$

4.2.1. Calculate the geometric mean for the following data;

Marks(x)	100-104	105-109	110-114	115-119	120-124
No. of students(f)	24	30	45	65	72

Solution

x	f	Mid point(m)	Log m	f log m
100-104	24	102	2.0086	48.20
105-109	30	107	2.0293	60.87
110-114	45	112	2.0492	92.23
115-119	65	117	2.0681	134.43
120-124	72	122	2.0863	150.21
	∑f=236			∑f log m=485.94

$$G.M = \text{Antilog} \frac{\sum f \log m}{\sum f}$$

=AL (485.94)\23

=AL (2.0590)

G.M = 114.6

4.2.2. Calculate the geometric mean for the following data

No. of family	50-54	55-59	60-64	65-69
No. of children	8	12	18	26

Solution

X	f	Mid point(m)	Log m	Flog m
50-54	8	52	1.7160	13.72
55-59	12	57	1.7559	21.06
60-64	18	62	1.7923	32.26
65-69	26	67	1.8260	47.47
	∑f =64			∑flog m =114.51

$$G.M = \text{Antilog} \frac{\sum f \log m}{\sum f}$$

G.M = AL (114.51/ 64)

=AL (1.7892)

G.M =61.54

4.2.3. Calculate the geometric mean for the following data;

x	12-17	18-23	24-29	30-35	36-41
f	8	19	22	28	30

Solution

x	f	Mid point	Log m	flog m
12-17	18	14.5	1.1613	9.29
18-23	19	20.5	1.3117	24.92
24-29	22	26.5	1.4232	31.31
30-35	28	32.5	1.5118	42.31
36-41	30	38.5	1.5854	47.56
	∑f=107			∑flog m =155.41

$$G.M = \text{Antilog} \frac{\sum f \log m}{\sum f}$$

G. M =AL (155.41/107)

=AL (1.4524)

G. M =28.34

4.2.4. Calculate the geometric mean for the following data;

x	0-10	10-20	20-30	30-40	40-50
f	7	12	24	35	45

Solution

x	f	Mid point (m)	Log m	f log m
0-10	7	5	0.6989	4.88
10-20	12	15	1.1760	14.11
20-30	24	25	1.3979	35.54
30-40	35	35	1.5440	15.44
40-50	45	45	1.6532	11.57
	$\sum f=60$			$\sum f\log m=79.54$

$$G.M = \text{Antilog}\frac{\sum f\log m}{\sum f}$$

G M= AL (79.54 /60)

=AL (1.3256)

=21.16

4.2.5. Calculate the geometric mean for the following data

No. of leaves	10-19	20-29	30-39	40-49	50-59
No. of plant	12	19	31	27	16

Solution

x	f	Mid point(m)	Log m	flog m
10-19	12	14.5	1.1613	13.93
20-29	9	24.5	1.3891	26.39
30-39	31	34.5	1.5378	47.67
40-49	27	44.5	1.6483	44.50
50-59	16	54.5	1.7363	27.78
	$\sum f= 103$			$\sum f\log m=160.27$

$$G.M = \text{Antilog}\frac{\sum f\log m}{\sum f}$$

G M= AL (160.27 /103)

=AL (1.5560)

=35.98

4.2.6. Calculate the geometric mean for the following data;

Rice(x)	100-110	110-120	120-130	140-150	150-160
Yield(f)	26	30	42	60	72

Solution

x	f	Mid point(m)	Log m	f log m
100-110	26	105	2.0211	52.54
110-120	30	115	2.0606	61.82
120-130	42	125	2.0969	88.06
130-140	60	135	2.1303	127.81
140-150	72	145	2.1613	155.61
	$\sum$ f=230			$\sum$ flog m=485.83

$$G.M = \text{Antilog}\frac{\sum f \log m}{\sum f}$$

G M= AL (485.83 /230)

= AL(2.1123)

G M =129.5

4.2.7. Calculate geometric mean for the following data;

No of labours	12-15	16-19	20-23	24-27	28-31
income	6	8	12	26	22

Solution

Labours	Frequency	Mid point	Log m	Flog m
12-15	6	13.5	1.1303	6.78
16-19	8	11.5	1.2430	9.94
20-23	12	21.5	1.3324	15.98
24-27	26	25.5	1.4065	36.57
28-31	22	29.5	1.4698	32.33
	$\sum$f=74			$\sum$ flog m =101.6

$$G.M = \text{Antilog}\frac{\sum f \log m}{\sum f}$$

G M= AL (101.6 /74)

= AL(1.3729)

=23.59

4.2.8. Calculate the geometric mean for the following data;

Yield	100-105	105-110	110-115	115-120	120-125
No. of grain	12	16	28	32	48

Solution

Yield	frequency	Mid point	Log m	Flog m
100-105	12	102.5	2.0107	24.12
105-110	16	107.5	2.0314	32.50
110-115	28	112.5	2.0511	57.43
120-125	32	117.5	2.0700	66.24
125-130	48	122.5	2.0881	100.23
	$\sum f$=136			$\sum$ flog m =280.52

$$G.M = \text{Antilog}\frac{\sum f \log m}{\sum f}$$

G .M =AL (280.52 /136)

=AL(2.0626)

G .M = 115.5

4.2.9 Merits and demerits of Geometric Mean

a) It is based upon all values of the given data.

b) It is capable of further mathematical treatment.

c) It is not much affected by sampling fluctuations.

Demerits of Geometric Mean

a) It is not easy to understand & not easy to calculate

b) It is not well defined.

c) If anyone data value is zero then GM is zero.

d) It cannot be calculated if any observations are missing.

e) It cannot be calculated for the data with open end classes.

f) It is affected by extreme values.

g) It cannot be located graphically.

h) It may be number which is not present in the data.

i) It cannot be calculated for the data representing qualitative characteristic

5.0 Harmonic Mean

5.1. Definition

The simple harmonic mean of a set of observation is defined as the reciprocal of the A.M. of the reciprocals of the observations.

Note: For any set of postitive numbers, it is always true that

harmonic means geometric means arithmetic means root mean square

5.1.1 Raw Data

If x1 , x2, …, xn are data values then harmonic mean is given by

$$HM = \frac{n}{\frac{1}{x1} + \frac{1}{x2} + \ldots + \frac{1}{xn}} = \frac{n}{\sum_{i=1}^{n} \frac{1}{xi}}$$

5.1.2.Ungrouped Frequency Distribution

If x1 , x2, …, xn are data values and f1 , f2, …, fn are the frequencies then harmonic mean is given by

$$HM = \frac{N}{\frac{f1}{x1} + \frac{f2}{x2} + \ldots + \frac{fn}{xn}} = \frac{N}{\sum \frac{fi}{xi}}$$

5.1.3. Frequency Distribution

If x1 , x2, …, xn are class marks or mid values and f1 , f2, …, fn are the frequencies then harmonic mean is given by

$$HM = \frac{N}{\frac{f1}{x1} + \frac{f2}{x2} + \ldots + \frac{fn}{xn}} = \frac{N}{\sum \frac{fi}{xi}}$$

where $N = \sum_{i=1}^{n} fi$

Example: For the numbers 4 and 9,

$$\text{harmonic mean} = \frac{2}{\frac{1}{4}+\frac{1}{9}} = \frac{72}{13} = 5.54$$

$$\text{geometric mean} = \sqrt{4.9} = \sqrt{36} = 6$$

$$\text{arithmetic mean} = \frac{4+9}{2} = 6.5$$

$$\text{root mean square} = \sqrt{\frac{4^2+9^2}{2}} = \sqrt{48.5} = 6.96$$

$$\text{Arithemtic mean} = \frac{x_1 + x_3 + \ldots + x_{y3}}{n}$$

$$\text{Geometric mean} = \sqrt[n]{x_1 \cdot x_3 \cdots x_{y3}}$$

$$\text{Harmonic mean} = \frac{n}{\frac{1}{x_1}+\frac{1}{x_2}+\cdots+\frac{1}{x_n}}$$

5.1.4. Merits and demerits of Harmonic Mean

Merits of Harmonic Mean

a) It is rigidly defined.
b) It is easy to understand & easy to calculate.
c) It is based upon all values of the given data.
d) It is capable of further mathematical treatment.
e) It is not much affected by sampling fluctuations.

Demerits of Harmonic Mean

a) It is not easy to understand & not easy to calculate.
b) It cannot be calculated if any observations are missing.
c) It cannot be calculated for the data with open end classes.
d) It is usually not a good representative of the data.
e) It is affected by extreme values
f) It cannot be located graphically.
g) It may be number which is not present in the data.
h) It can be calculated for the data representing qualitative characteristic.

2

Measures of Dispersion

2.1. Introduction

Measures of dispersion or variability are used to describe the spread or dispersion of the data. Naturally, Dispersion is the extent to which values in a distribution differ from the average of the distribution.

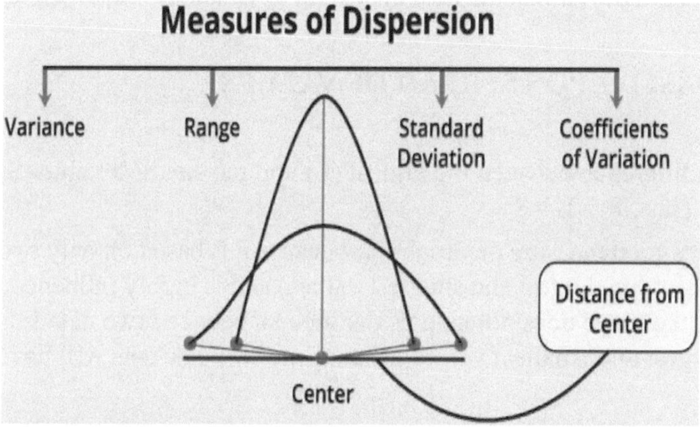

2.2. Measures of dispersion can be classified into two categories shown below

1. Absolute Measures of Dispersion
2. Relative Measures of Dispersion

2.3. Characteristics of a Good Measure of Dispersion

- It should be easy to calculate and simple to understand.
- It should be based on all the observations of the series.
- It should be rigidly defined.
- It should not be affected by extreme values.
- It should not be unduly affected by sampling fluctuations.
- It should be capable of further mathematical treatment and statistical analysis.

2.4. Absolute Measures of Dispersion

These measures of dispersion are measured and expressed in the units of data themselves. For example-Meters, Dollars, kilograms, centimeters etc. Some absolute measures of dispersion are:

1. Range
2. Quartile Deviation
3. Mean Deviation
4. Standard Deviation

Apart from these measures which give a numerical value, there is a graphic method for estimating dispersion. Range and quartile deviation measure the dispersion by calculating the spread within which the values lie. Mean deviation and standard deviation calculate the extent to which the values differ from the average.

MEASURES BASED UPON SPREAD OF VALUES

2.4.0. Range

Range (R) is the difference between the largest (L) and the smallest value (S) in a distribution. Thus, R = L – S

Range is not a very good measure of variability because it is based on only two values in the data set (the largest and smallest values) and is highly influenced by outliers. Also, the range does not help us distinguish between two data sets with the same largest and smallest values because the two data sets will have the same range.

2.4.1. Find out the range, Coefficient of range of the weight of rice in kilogram. 5, 8, 3, 10, 23, 4.

Solution

5, 8, 3, 10, 23, 4.

L- S=23-3

Range=20

$$\text{Coefficient of range} = \frac{L-S}{L+S} = .76$$

2.4.2. find the range, Coefficient of range of marks in a student 96, 54, 45, 76, 80

Solution

96, 54, 45, 76, 80

L- S = 96-45

Range =51

$$\text{Coefficient of range} = \frac{L-S}{L+S} = .36$$

2.4.3. Find the range of the given data 76, 45, 34, 23, 10.

solution

76, 45, 34, 23, 10,

= 76-10

= 66

2.4.4. Find the range of following data 3, 5, 4, 7, 6.

Solution

3, 5, 4, 7, 6.

= 6-3

= 3

2.4.5. Find the range of the following data 123, 145, 164, 100, 120

Solution

123, 145, 164, 100, 120

= 164-100

= 64

DISCRETE TYPE

2.4.6. Find out the range of given data

No .of student	56	34	85	60
marks	69	56	87	99

Solution

= 85-34

= 51

2.4.7. Find the range of the following data

Hill	5	8	7	3	9
No. of plant	15	6	13	8	12

Solution

= 9-3

= 6

2.4.8. Find the range of yield of rice plant

Rice plant	15	23	17	21	10
Yield	5	6	9	3	6

Solution

= 23-10

= 13

2.4.9. Find the range of following data

Red gram plant	45	34	48	76	41
No. of seed	10	68	69	143	55

Solution

= 76-34

= 42

2.4.10. Find the range of the following data

No. of labour	5	4	7	2	1
Cost	25	21	35	12	6

Solution

= 7-1

= 6

2.5. QUARTILE DEVIATION

Quartile: Its divide the list of numbers or data into quarters.

2.5.0. Quartile Deviation: Quartile Deviation is the measure of the difference between the upper and lower quartile. This measure of deviation is also known as the interquartile range.

2.5.1. Calculate the quartile deviation and its coefficient from the following data.

Roll no.	1	2	3	4	5	6	7	8	9	10	11	12
Marks	39	40	40	41	41	42	42	43	43	44	44	45

Solution

Computation of quartile deviation and its coefficient

Roll no.	Marks
1	39
2	40
3	40
4	41
5	41
6	42
7	42
8	43
9	43
10	44
11	44
12	45

Quartile Deviation (or) Q.D. $=\frac{Q_3-Q_1}{2}$

$Q1=\text{Size of }\frac{N+1}{4}^{th}\text{ item}=\frac{12+1}{4}=3.25\text{ th item}$

Size of 3.25 th item=3 rd item +.25(4th - 3rd item)

=40+.25(41-40)=40.25

$Q_3=\text{Size of }3\ \frac{(N+1)}{4}^{th}\text{ item}=3\frac{(12+1)}{4}$

=9.75 th item

Size of 9.75 item=9 th item +.75(10th - 9th item)

=43+.75(44-43)

=43+.75=43.75

$$Q.D.=\frac{43.75-40.25}{2}$$

$$=\frac{3.5}{2}=1.75$$

Coefficient of Q.D.$=\frac{Q_3-Q_1}{Q_3+Q_1}$

$$= \frac{43.75 - 40.25}{43.75 + 40.25}$$

$$= \frac{3.5}{84} = .042$$

RESULT

Quartile Deviation =1.75

co-efficient of **Quartile** Deviation =0.042

2.5.2. Calculate the quartile deviation and its coefficient from the following data.

Height in inches	58	59	60	61	62	63	64	65	66
No.of students	15	20	32	35	33	22	20	10	8

Solution

Height in inches	**f**	**c.f**
58	15	15
59	20	35
60	32	67
61	35	102
62	33	135
63	22	157
64	20	177
65	10	187
	N=Σf=195	

Q_1 = Size of $\frac{N+1}{4}^{th}$ item = $\frac{195+1}{4}$ = 49 th item

Size of 49 th item = 60

$\therefore$ **Q_1=60**

Q_3 = Size of $3\frac{(N+1)}{4}$ th item = $3\frac{(195+1)}{4}$

=147th item

Size of 147 th item is 63

$\therefore$ **Q_3**= 63

$$Q.D. = \frac{Q_3 - Q_1}{2}$$

$$= \frac{63-60}{2}$$

Q.D. = 1.5

$$\text{Coefficient of Q.D.} = \frac{Q_3 - Q_1}{Q_3 + Q_1}$$

$$= \frac{63-61}{63+61}$$

= 0. 012

RESULT

Quartile Deviation =1.5

co-efficient of **Quartile** Deviation =0.012

2.5.3. calculate co-efficient of Quartile Deviation from the following data.

Marks	4-8	8-12	12-16	16-20	20-24	24-28	28-32	32-36	36-40
Students	6	10	18	30	15	12	10	6	2

Solution

Marks	**f**	**c.f**
4-8	6	6
8-12	10	16
12-16	18	34
16-20	30	64
20-24	15	79
24-28	12	91
28-32	10	101
32-36	6	107
36-40	2	109
	N=Σf=109	

$$\text{Coefficient of Q.D.} = \frac{Q_3 - Q_1}{Q_3 + Q_1}$$

$$Q_1 = \text{Size of N/4}^{th} \text{ item} = \frac{109}{4}$$

= 27.25 th item

Q_1 lies in the class 12-16

$$Q1 = L + \frac{N/4\text{-c.f}}{f} \times c$$

L= 12,N/4=27.5, c.f. =16,f= 18,c=4

$$Q_1 = \frac{12 + 27.5\text{-}16}{18} \times 4$$

$$=12+\frac{11.25}{18} \times 4$$

=12+2.5=14.5

$\therefore$ **Q_1**=14.5

$$Q_3 = \text{Size of } 3N/4^{th} \text{ item} = \frac{3x109}{4}$$

= 81.75th item

Q_3 lies in the class 24-28

$$Q_3 = \frac{L + 3N/4\text{-c.f}}{f} \times c$$

L= 24, 3N/4=81.75, c.f. =79,f= 12,c=4

$$Q_3 = 24 + \frac{(81.75\text{-}79)}{12} \times 4$$

$$=24+\frac{11}{12} \times 4$$

=24+.917=24.917

$\therefore$ **Q_1**=14.5, $\therefore$ **Q_3**=24.917

$$\text{Coefficient of Q.D.} = \frac{24.917\text{-}14.50}{24.917+14.50}$$

$$=\frac{10.417}{39.417}$$

=0.264

RESULT

Co-efficient of **Quartile** Deviation =0.264

Mean Deviation

Individual observations

$$\text{M.D.} = \frac{\sum /D/}{N}$$

/D/= Deviations from median or mean

2.5.4. Calculate mean deviation from(i)mean (ii) mode (iii) median in respect of the marks obtained by five students given below.4,5,6,4.5,5.

Solution: calculation of mean deviation from mean, mode, median.

$$\text{Mean:} = \bar{X}\frac{4+5+6+4.5+5}{5}$$

$$= 4.9$$

Mode: 5(since 5 is repeated the maximum number of times i.e. 5)

Median:4,4.5,5,5,6.

Computation of mean deviation

Marks x	Deviations from mean |D|	Deviations from mode |D|	Deviations from median |D|
4	0.9	1	1
5	0.1	0	0
6	1.1	1	1
4.5	0.4	0.5	0.5
5	0.1	0	0
ΣX= 24.5	Σ|D|=2.6	Σ|D|=2.5	Σ|D|= 2.5

$$\text{Mean deviation (from mean)} = \frac{\sum /D/}{N} = 2.6/5 = 0.52$$

$$\text{Mean deviation (from mean)} = \frac{\sum /D/}{N} = 2.5/5 = 0.5$$

$$\text{Mean deviation (from mean)} = \frac{\sum /D/}{N} = 2.5/5 = 0.5$$

2.5.5. Calculate mean deviation from (i)mean (ii) mode (iii) median for the following data 8, 5, 6, 6, 5.5.

Solution: calculation of mean deviation from mean, mode, median.

$$\text{Mean: } \bar{X} = \frac{8+5+6+6+5.5}{5}$$

$$= 6.1$$

Mode: 6(since 6 is repeated the maximum number of times i.e. 6)

Median:5, 5.5, 6, 6,8.

Computation of mean deviation

Marks x	Deviations from mean |D|	Deviations from mode |D|	Deviations from median |D|
8	1.9	2	2
5	1.1	1	1
6	0.1	0	0
6	0.1	0	0
5.5	0.6	0.5	0.5
ΣX=30.5	Σ|D|=3.8	Σ|D|=3.5	Σ|D|= 3.5

Mean deviation (from mean) = $\frac{\sum f|D|}{N}$ =3.8/5= 0.76

Mean deviation (from median) = $\frac{\sum f|D|}{N}$ = 3.5/5 = 0.7

Mean deviation (from mode) = $\frac{\sum f|D|}{N}$ = 3.5/5 = 0.7

2.5.6. calculate the mean deviation from (i)mean (ii) mode (iii) median in respect of marks obtained by nine students given below and show that the mean deviation from median is minimum. Marks (out of 25)7,4,10,9,15,12,7,9,7.If the marks are doubled (converted out of 50) will the variation in marks increase? Give reasons.

Solution: calculation of mean deviation from mean, mode, median.

Mean: $\bar{X} = \frac{7+4+10+9+15+12+7+9+7}{5}$

= 80/9 = 8.89

Mode: 7(since 7 is repeated the maximum number of times i.e. 7)

Median: 9

computation of mean deviation

Marks x	Deviations from mean |D|	Deviations from median |D|	Deviations from mode |D|
7	1.9	2	0
4	4.9	5	3
10	1.1	1	3
9	0.1	0	2
15	6.1	6	8
12	3.1	3	5
7	1.9	2	0
9	0.1	0	2
7	1.9	2	0
ΣX=80	Σ|D|=21.1	Σ|D|=21	Σ|D|= 23

Mean deviation (from mean) $= \frac{\sum |D|}{N} = 21.1/9 = 2.34$

Mean deviation (from median) $= \frac{\sum |D|}{N} = 21/9 = 2.33$

Mean deviation (from median) $= \frac{\sum |D|}{N} = 23/9 = 2.56$

From these calculation it is clear that the mean deviation is least from median. When the marks are doubled , i.e., converted out of 50 the variation in marks would also increase and in fact it would be doubled. **The reason is that when all the values of a series are multiplied by a certain constant**, say two , the value of mean ,median , mode, mean deviation , standard deviation etc. would also be increased and shall be obtained by multiplying with that constant.

2.6. Discrete and continuous series

$$\text{M.D.} = \frac{\sum f|D|}{N}$$

In case mean and median is in fraction the above formula will involve difficulties of computations in that case the following formula is applied:

$$\frac{\sum fma - \sum fam - \left(\sum fa - \sum fb\right) x\text{Median}}{N}$$

$\sum fm_a$ and $\sum fm_b$ stand for totals of products of midpoints and frequencies corresponding to midpoints above and below the average value respectively.

$\sum f_a$ and $\sum f_b$ stand for the totals of frequencies pertaining to midpoints above and below the average value.

In case of deviations are taken from mean than instead median mean should be put in the above formula.

$$\text{Coeff. Of M.D.} = \frac{\text{M.D}}{\text{Median}}$$

Note

In case mean deviation is computed from mean, then coefficient of mean deviation will be computed by dividing mean deviation by mean.

2.6.1. Calculate mean deviation from (taking deviations from mean) from the following data

x	5	10	15	20	25
f	7	4	3	4	7

Solution

Computation of mean deviation

x	f	fx	D=(x- mean)	/D/	f/D/
5	7	35	-10	10	70
10	4	40	-5	5	20
15	3	45	0	0	3
20	4	80	5	5	20
25	7	175	10	10	70
	Σf=N=25	Σfx=375			Σf/D/=183

$$\bar{X} = \frac{\sum fx}{\sum f}$$

$= 375/25 = 15$

$$\text{M.D} = \frac{\sum f\,|D|}{N}$$

$= 183/25$

M.D.= 7.32

$$\text{Mean co-efficient of dispersion} = \frac{\text{Mean deviation}}{\text{Mean}}$$

$= 7.32/15$

$= 0.488$

RESULT

Mean Deviation =7.32

Mean co-efficient of dispersion =0.488

2.6.2. Calculate the mean deviation and co-efficient of dispersion from the following data.

MARKS	10	15	20	30	40	50
FREQUENCY	8	12	15	10	3	2

Solution

Computation of mean deviation

x	f	fx	/D/	f/D/
10	8	80	11.6	92.8
15	12	180	6.6	79.2
20	15	300	1.6	24
30	10	300	8.4	84
40	3	120	18.4	55.2
50	2	100	28.4	56.8
	N= Σf=50	Σfx=1080		Σf/D/=392

$$\bar{X} = \frac{\sum fx}{\sum f}$$

= 1080/50

= 21.6

$$M.D = \frac{\sum f|D|}{N}$$

=392/50

M.D. =7.84

$$\text{Mean Co-efficient of dispersion} = \frac{\text{Mean deviation}}{\text{Mean}}$$

= 7.84/21.6

= 0.363

RESULT

Mean Deviation =7.84

Mean co-efficient of dispersion =0.363

2.6.3. Calculate the mean deviation from the following data.

x	2	4	6	8	10
f	1	4	6	4	1

Solution

Computation of mean deviation

x	f	fx	D=(x-x̄)	/D/	f/D/
2	1	2	-4	4	4
4	4	16	-2	2	8
6	6	36	0	0	6
8	4	32	2	2	8
10	1	10	4	4	4
	N =Σf =16	Σfx=96	0		Σf/D/=24

$$\bar{X} = \frac{\sum fx}{\sum f}$$

=96/16 = **6**

$$M.D. = \frac{\sum f|D|}{N}$$

=24/16

M.D.=1.5

$$\text{Mean co-efficient of dispersion} = \frac{\text{Mean Deviation}}{\text{Mean}}$$

=1.5/6

=0.25

RESULT

Mean Deviation =1.5

Mean co-efficient of dispersion =0.25

2.6.4. Calculate the mean deviation from the following data.

Marks	50	51	52	53	54
No.of students	2	3	4	1	2

Solution

Computation of mean deviation

x	f	fx	D=(x-x̄)	/D/	f/D/
50	2	100	-1.8	1.8	3.6
51	3	153	-0.8	0.8	2.4
52	4	208	0.2	0.2	0.8
53	1	53	1.2	1.2	1.2
54	2	108	2.2	2.2	4.4
	N=Σf=12	Σfx=622			Σf/D/=12.4

$$\bar{X} = \frac{\sum fx}{\sum f}$$

=622/12

=51.8

$$\text{M.D.} = \frac{\sum f|D|}{N}$$

=12.4/12

=1.03

$$\text{Mean co-efficient of dispersion} = \frac{\text{Mean Deviation}}{\text{Mean}}$$

=1.03/51.8

= 0.019

RESULT

Mean Deviation=1.03

Mean co-efficient of dispersion=0.019

2.6.5. Calculate the mean deviation from the following data.

Size of item	3-4	4-5	5-6	6-7	7-8	8-9	9-10
Frequency	3	7	22	60	85	32	8

Solution

Computation of mean deviation

Size of item	f	Midpoint(m)	m – 6.5 =d	fd	fm
3-4	3	3.5	3.5-6.5= -3	-9	10.5
4-5	7	4.5	-2	-14	31.5
5-6	22	5.5	-1	-22	121.0
6-7	60	6.5	0	0	390.0
7-8	85	7.5	1	85	637.5
8-9	32	8.5	2	64	272.0
9-10	8	9.5	3	24	76.0
	N=∑f=217			∑fd=128	

$$\text{M.D.} = \frac{\sum f\,|D|}{N}$$

/D/ denotes deviations from mean (ignoring signs)

$$\bar{X} = A + \frac{\sum fd}{\sum f}$$

A=6.5, Σfd=128,N=217

$$\bar{X} = 6.5 + \frac{128}{217} = 6.5+0.59 = 7.09$$

Since **mean is in fraction** it would be to apply the following formula for calculating mean deviation.

$$\text{M.D.} = \frac{\sum \text{fma} - \sum \text{fma} - (\sum \text{fa} - \sum \text{fb})\text{xMEAN}}{N}$$

$\sum fm_a = 637.5+272.0+76.0 = 985.5$

$\sum f_a = 85+32+8 = 125$

$\sum fm_b = 10.5+31.5+121+390 = 553$

$\sum f_b = 3+7+22+60 = 92$

N= 217

$$M.D = \frac{985.5 - 533 - \{(125 - 92) \times 7.09\}}{217}$$

$$= \frac{985.5 - 533 - 233.97}{217}$$

$$M.D. = \frac{218.53}{217}$$

RESULT

M.D.=1.007

2.6.6. Calculate the mean deviation from the following data.

Size of item	1-2	2-3	3-4	4-5	5-6
Frequency	25	12	18	9	32

Solution

Computation of mean deviation

Size of item	Frequency f	Midpoint (m)	m –x=d m-3.5=d	fd	fm
1-2	25	1.5	-2	-50	37.5
2-3	12	2.5	-1	-12	30.0
3-4	18	3.5	0	0	63
4-5	9	4.5	1	9	40.5
5-6	32	5.5	2	64	176
	N=∑f=96			Σfd=11	

$$M.D. = \frac{\sum f\,|D|}{N}$$

/D/ denotes deviations from mean (ignoring signs)

$$\bar{X} = A + \frac{\sum fd}{\sum f}$$

A=3.5, Σfd=11, N=∑f =96

$$\bar{X} = 3.5 + \frac{11}{96} = 3.5 + 0.11 = 3.61$$

Since **mean is in fraction** it would be to apply the following formula for calculating mean deviation.

$$M.D. = \frac{\sum fma - \sum fmb - \{(\sum fa - \sum fb) \times Mean\}}{N}$$

$\sum fm_a = 40.5 + 176 = 216.5$

$\sum f_a = 9+32=41$

$\sum fm_b = 37.5+30.0+63 = 130.5$

$\sum f_b = 25+12+18 = 55$

N = 96

$$\text{M.D.} = \frac{216.5-130.5-\{(41-55)\times 3.61\}}{96}$$

$$= \frac{86+50.54}{96}$$

$$\text{M.D.} = \frac{136.54}{96}$$

M.D.=1.422

RESULT

MEAN DEVIATION = 1.422

2.7. Standard Deviation

The standard deviation is defined as the square root of the mean of the squared deviations of individual values from their mean and it is denoted by σ

$$\text{SD} = \sqrt{\frac{\sum d^2}{n}}$$

$d = X - \bar{X}$

The sample standard deviation should be an unbiased estimate of the population standard deviation.

S.D. For raw data

2.7.1. Calculate the S.D. and Co-efficient of variance following data. (Taking deviations from actual mean) 6,15,20,18,11

Solution

X	Mean ($\bar{X}$)	$d=X-\bar{X}$	d^2
6	14	-8	64
15	14	1	1
20	14	6	36
18	14	4	16
11	14	-3	9
n=5			$\sum d^2 = 126$

$$\text{Mean}(\bar{X}) = \frac{\text{sum of observation}}{\text{no. of .observation}}$$

$$= \frac{70}{5} = 14$$

$\sum d^2 = 126$

$$\text{Standard deviation} = \text{SD} = \sigma = \frac{\sqrt{\sum d^2}}{n}$$

$$= \sqrt{\frac{126}{5}}$$

=5.O199

$$\text{Co-efficient of variance} = \frac{\text{S.D.}}{\bar{X}} \times 100$$

$$= \frac{5.01}{14} \times 100$$

= 35.79

RESULT

Standard deviation =5.0199

Co-efficient of variance =35.79

2.7.2. Calculate the S.D. and Co-efficient of variance following data. (Taking deviations from assumed mean 15) 6,15,20,18,11

Solution

Taking deviations from 15

X	Mean($\bar{X}$)	d=X- $\bar{X}$	d^2
6	15	-9	81
15	15	0	0
20	15	5	25
18	15	3	9
11	15	-4	16
n=5		$\sum d = -5$	$\sum d^2 = 131$

$$\text{Standard deviation} = \text{SD} = \sigma = \sqrt{\frac{\sum d^2}{n}\left(\left(\frac{\sum d^2}{n}\right)^2\right)}$$

$$= \sqrt{\frac{131}{5} - \left(\left(\frac{-5}{5}\right)^2\right)}$$

=5.0199

$$\text{Co-efficient of variance} = \frac{\text{S.D}}{\overline{X}} \times 100$$

$$= \frac{5.0199}{15} \times 100$$

= 33.46

RESULT

Standard deviation =5.0199

Co-efficient of variance =33.46

2.7.3. Calculate the S.D. and Co-efficient of variance following data. (Step deviation method) 6,15,20,18,11

Solution

(Step deviation method)

X	Mean($\overline{X}$)	d=X- $\overline{X}$	d=X- $\overline{X}$ /5	d^2
6	15	-9	-1.8	3.24
15	15	0	0	0
20	15	5	1	1
18	15	3	0.6	0.36
11	15	-4	-0.8	0.64
n=5			$\sum d$=-1	$\sum d^2$ =5.24

$$\text{Standard deviation} = \text{SD} = \sigma = \sqrt{\frac{\sum d^2}{n} - \left(\left(\frac{\sum d^2}{n}\right)\right)} \text{xi}$$

$$= \sqrt{\frac{5.24}{5} - \left(\left(\frac{-1}{5}\right)^2\right)} \text{x5}$$

=5.0199

$$\text{Co-efficient of variance} = \frac{\text{S.D.}}{\overline{X}} \times 100$$

$$= \frac{5.0199}{15} \times 100$$

= 33.46

RESULT

Standard deviation =5.0199

Co-efficient of variance =33.46

2.7.4. Calculate the S.D. from the following data. (Taking deviations from actual mean) 600,620,640,620,680,670,680,640,700,650.

Solution

X	**Mean($\overline{X}$)**	**d=X-$\overline{X}$**	**d^2**
600	650	-50	2500
620	650	-30	900
640	650	-10	100
620	650	-30	900
680	650	30	900
670	650	20	400
680	650	30	900
640	650	-10	100
700	650	50	2500
650	650	0	0
$\sum x$=6500			$\sum d^2$ =9200

$$\text{Mean}(\overline{X}) = \frac{\text{sum of observation}}{\text{no. of .observation}}$$

$$= \frac{6500}{10} = 650$$

$\sum d^2$ =9200

$$\text{Standard deviation} = \text{SD} = \sigma = \sqrt{\frac{\sum d2}{n}}$$

$$= \frac{\sqrt{9200}}{10}$$

=30.33

$$\text{Co-efficient of variance} = \frac{\text{S.D.}}{\overline{X}} \times 100$$

$$= \frac{30.33}{650} \times 100$$

= 4.66

RESULT

Standard deviation =30.33

Co-efficient of variance =4.66

2.7.5. Calculate the S.D. and Co-efficient of variance following data. (Taking deviations from assumed mean 640) 600,620,640,620,680,670 ,680,640,700,650.

Solution

Taking deviations from 640

X	**Mean($\overline{X}$)**	**d=X- $\overline{X}$**	**d^2**
600	640	-40	1600
620	640	-20	400
640	640	0	0
620	640	-20	400
680	640	40	1600
670	640	30	900
680	640	40	1600
640	640	0	0
700	640	60	3600
650	640	10	100
$\sum x$=6500		$\sum d$=100	$\sum d^2$ =10,200

$$\text{Standard deviation} = \text{SD} = \sigma = \sqrt{\frac{\sum d^2}{n} - \left(\left(\frac{\sum d^2}{n}\right)^2\right)}$$

$\sum d^2$ =10,200,$\sum d$=100,n= 10

$$\sigma = \sqrt{\frac{10,200}{10} - \left(\left(\frac{100}{10}\right)^2\right)}$$

=30.33

Co-efficient of variance $= \frac{S.D.}{\overline{X}} \times 100$

$= \frac{30.33}{640} \times 100$

= 4.73

RESULT

Standard deviation =30.33

Co-efficient of variance =4.73

2.7.6. Calculate the S.D. and Co-efficient of variance following data. (Step deviation method) 6,15,20,18,11

Solution

(Step deviation method)

X	Mean($\overline{X}$)	d=X- $\overline{X}$	d=X- $\overline{X}$ /10	d^2
600	640	-40	-4	16
620	640	-20	-2	4
640	640	0	0	0
620	640	-20	-2	4
680	640	40	4	16
670	640	30	3	9
680	640	40	4	16
640	640	0	0	0
700	640	60	6	36
650	640	10	1	1
n=10			$\sum d=10$	$\sum d^2 = 102$

Standard deviation = SD = $\sigma = \sqrt{\frac{\sum d^2}{n} - \left(\left(\frac{\sum d}{n}\right)^2\right)}$ xi

$\sum d^2 = 102, \sum d = 10, n = 10$

$= \sqrt{\frac{102}{10} - \left(\left(\frac{10}{10}\right)^2\right)}$ x10

$\sigma = 30.33$

$$\text{Co-efficient of variance} = \frac{\text{D.S.}}{\bar{X}} \times 100$$

$$= \frac{30.33}{640} \times 100$$

= 4.73

RESULT

Standard deviation =30.33

Co-efficient of variance =4.73

S.D. For discrete frequency distribution

2.7.7. Calculate S.D. and Co-efficient of variance from the following data

x	35	40	45	50	55
f	2	3	5	3	2

Solution

x	f	fx	$\bar{X}$	d=X- $\bar{X}$	d^2	fd^2
35	2	70	45	-10	100	200
40	3	120	45	-5	25	75
45	5	225	45	0	0	0
50	3	150	45	5	25	75
55	2	110	45	10	100	200
		$\sum$ fx=675				$\sum fd^2 = 550$

$\bar{X} = \sum fx / \sum f$

= 675/15=45 (integer)

$$(\text{S.D})\sigma = \sqrt{\frac{\sum fd^2}{\sum f}}$$

$$= \sqrt{\frac{550}{15}}$$

$$= \sqrt{36.67}$$

$$= 6.06$$

Co-efficient of variance $= \frac{S.D.}{\bar{X}} \times 100$

$= \frac{6.06}{45} \times 100$

$= 13.46$

RESULT

Standard deviation =6.06

Variance=(S.D)^2 =36.67

Co-effcient of Variance =13.46

2.7.8. Calculate S.D., variance and Co-efficient of variance from the following data

x	10	20	30	40	50	60
f	8	12	20	10	7	3

Solution

x	f	fx	$\bar{X}$	d=X-$\bar{X}$	fd	d^2	fd^2
10	8	80	30	-20	-160	400	3200
20	12	240	30	-10	-120	100	1200
30	20	600	30	0	0	0	0
40	10	400	30	10	100	100	1000
50	7	350	30	20	140	400	2800
60	3	180	30	30	90	900	2700
	n =∑f=60	∑fx=1850			∑ fd=50		∑ fd^2 = 10,900

$\bar{X} = \sum fx / \sum f$

=1850/60

=30.83 (not an integer)

$$\sigma = \sqrt{\frac{\sum fd^2}{n} - \left(\left(\frac{\sum fd}{n}\right)^2\right)}$$

$$= \sqrt{\frac{109000}{60} - \left(\left(\frac{50}{60}\right)^2\right)}$$

$= \sqrt{18166 - .69.}$

$= \sqrt{180.97}$

=13.45

$$\text{Co-efficient of variance} = \frac{\text{S.D.}}{\overline{X}} \times 100$$

$$= \frac{17.90}{45} \times 100$$

=43.63

RESULT
Standard deviation =13.45
variance = $(S.D)^2$ = 180.97
Co-efficient of variance =43.63

2.7.9. Calculate S.D. and Co-efficient of variance from the following data

x	15	30	45	60	25
f	2	4	6	8	5

Solution

x	**f**	**fx**	$\overline{X}$	**d=X-** $\overline{X}$	d^2	fd^2
15	2	30	41	-26	676	1352
30	4	120	41	-11	121	484
45	6	270	41	4	16	96
60	8	480	41	20	400	4800
25	5	125	41	-16	256	1280
	$\sum f=25$	$\sum fx=1025$				$\sum fd^2 = 8012$

$\overline{X}$= 1025 /25 =41(integer)

$$(\text{S.D.}) = \sigma = \sqrt{\frac{\sum fd^2}{\sum f}}$$

$$= \sqrt{\frac{8012}{25}}$$

$$= \sqrt{320.48}$$

= 17.90

$$\text{Co-efficient of variance} = \frac{\text{S.D.}}{\overline{X}} \times 100$$

$$= \frac{17.90}{45} \times 100$$

= 39.77

RESULT
Standard deviation =17.90
Variance = $(S.D)^2$ = 320.48
Co-efficient of variance =39.77

2.7.10. Calculate S.D., variance and Co-efficient of variance from the following data

x	10	12	17	21	26
f	2	4	8	5	1

Solution

x	f	fx	$\bar{X}$	d=X- $\bar{X}$	fd	d^2	fd^2
10	2	20	17	-7	-14	49	98
12	4	48	17	-5	-20	25	100
17	8	136	17	0	0	0	0
21	5	105	17	4	20	16	80
26	1	26	17	9	9	81	81
	n =∑f=20	∑fx=335			∑ fd=-5		$\sum fd^2$ = 359

$\bar{X} = \sum fx / \sum f$

=335/20

=16.75 (not an integer)

$$\sigma = \sqrt{\frac{\sum fd^2}{n} - \left(\left(\frac{\sum fd}{n}\right)^2\right)}$$

$$= \sqrt{\frac{359}{20} - \left(\left(\frac{-5}{20}\right)^2\right)}$$

$$= \sqrt{17.95 - 0.625}$$

$$= \sqrt{17.8875}$$

=4.23

variance = $(S.D)^2$=17.88

$$\text{Co-efficient of variance} = \frac{S.D.}{\bar{X}} \times 100$$

$$= \frac{4.23}{16.75} \times 100$$

=25.25

RESULT	
Standard deviation	=4.23
variance = (S.D)^2	= 17.88
Co-efficient of variance	=25.25

2.7.11. Calculate S.D. from the following data

x	5-25	25-45	45-65	65-85	95-105	105-125	125-145	145-165
f	4	6	14	22	14	5	7	3

Solution

x	f	m	$\bar{X}$	d=X- $\bar{X}$ /20	fd	d^2	fd^2
5-25	4	15	75	-3	-12	9	36
25-45	6	35	75	-2	-12	4	24
45-65	14	55	75	-1	-14	1	14
65-85	22	75	75	0	0	0	0
95-105	14	95	75	1	14	1	14
105-125	5	115	75	2	10	4	20
125-145	7	135	75	3	21	9	63
145-165	3	155	75	4	12	16	48
	n=$\sum$ f=75	$\sum$ fx=1025					$\sum$ fd^2 = 219

$$\bar{X} = A + \frac{\sum fd}{\sum f} xi$$

A= 75, $\sum$ fd =19, $\sum$ f=75, i=20

$$\bar{X} = 75 + \frac{19}{75} \text{ x } 20 = 75+5.07=80.07$$

$$\sigma = \sqrt{\frac{\sum fd^2}{n} - \left(\left(\frac{\sum fd}{n}\right)^2\right)} xi$$

$$= \sqrt{\frac{219}{75}\left(\left(\frac{19}{75}\right)^2\right)} \text{ x } 20$$

$$= \sqrt{292 - 0.006.^2} \text{ x } 20$$

1.69 x 20

=33.8

$$\text{Co-efficient of variance} = \frac{\text{S.D.}}{\bar{X}} \times 100$$

$$= \frac{33.8}{80.07} \times 100$$

$$= 42.21$$

RESULT
Standard deviation =33.8
Co-efficient of variance =42.21

2.7.12. Calculate S.D. and its coefficient from the following data

x	0-10	10-20	20-30	30-40	40-50
f	12	33	30	15	10

Solution

x	f	m	$\bar{X}$	d=X- $\bar{X}$ /10	fd	d^2	fd^2
0-10	12	5	25	-2	-24	4	48
10-20	33	15	25	-1	-33	1	33
20-30	30	25	25	0	0	0	0
30-40	15	35	25	1	15	1	15
40-50	10	45	25	2	20	4	40
	n=$\sum$ f=100				$\sum$fd=-22		$\sum fd^2 = 136$

$$\bar{X} = A + \frac{\sum fd}{\sum f} xi$$

A= 25, $\sum$ fd =-22, $\sum$ f=100, i=10

$$\bar{X} = 25 - \frac{22}{100} x10 = 25 - 2.2 = 22.8$$

$$\sigma = \sqrt{\frac{\sum fd^2}{n} - \left(\left(\frac{-22}{100}\right)\right)} x10$$

$$= \sqrt{\frac{136}{100} - \left(\left(\frac{-22}{100}\right)^2\right)} x10$$

$$= \sqrt{1.36 - 0.048}\, x10$$

=1.145 x 10

=11.45

Co-efficient of variance = $\frac{S.D.}{\overline{X}} \times 100$

$= \frac{11.45}{22.8} \times 100$

$= 0.502 \times 100$

$= 50.2$

Co-efficient of S.D. $= \frac{S.D.}{\overline{X}}$

$= \frac{11.45}{22.8}$

$= 0.502$

RESULT
Standard deviation =33.8
Co-efficient of variance =42.21
Co-efficient of S.D. = 0.502

2.8. For ungrouped data

$$s = \sqrt{\frac{\sum(x - \bar{x})^2}{n-1}}$$

For practical purpose

$$s = \frac{\sqrt{\sum x^2 - (\sum x)^2 / n}}{n-1}$$

2.9. For grouped data

$$s = \sqrt{\frac{\sum fx^2 - (\sum fx)^2 / n}{n-1}}$$

The term $\Sigma x^2 - (\Sigma x)^2/n$ is called the sum of squares and the variance is the

Mean square $= \frac{\sum x^2 - (\sum x)2/n}{n-1}$

Note: The standard deviation is denoted by **'s'** in case of sample and by **'σ'** in case of population.

2.10. Characteristics of SD

1. It is rigidly defined
2. Computation is based on all the observations
3. If all the variable values are same, then SD = 0
4. SD is least affected by fluctuations of sampling.

2.11. Uses

It is used in computing different statistical quantities and also used in testing the reliability of certain statistical measures.

2.12. Variance

square of the SD is called the variance i.e. variance = $(SD)^2$

The co-efficient of variation (CV) is given by

$$CV = \frac{SD}{Mean} \times 100$$

CV is independent of the units in which the mean and S.D are measured. The lesser the co-efficient of variation more consistent is the data. Thus to compare consistency of two data, we calculate their co-efficient of variation and say that the data with lesser co-efficient of variation is more consistent.

2.13. Characteristic of CV

The C.V is a **unit free measure.** It is always expressed as percentage. C.V will be small if the variation is small of the two groups, the one with less C.V is said to be more consistent.

2.14. Relative Measures of Dispersion

Relative Measure of Dispersion in Statistics are the values without units. A relative measure of dispersion is used to compare the distribution of two or more datasets.

The definition of the Relative Measure of Dispersion is the same as the Absolute Measure of Dispersion; the only difference is the measuring quantity. These measures of deviation are expressed in the form of ratios, percentages.

For example-Standard Deviation divided by the mean is an example of a relative measure. These measures are always dimensionless and are also known as the coefficient of dispersion. These measures come in handy while comparing the variation of two datasets that have different units. For example, consider two datasets of weights of students. In one dataset, the weight is measured in Kilograms, and in another one, it is measured in grams. Both will have equivalent variation in the values but since the units are different, absolute measures of dispersion will give a very high value for the dispersion in the dataset with weights in grams. Since absolute measures of dispersion are not appropriate in these cases, the relative measures of dispersion are used.

2.15. Lorenz Curve

The Lorenz curve is an important part of economics. It is a representation of the distribution of wealth and income. It was developed by Max.O. Lorenz to represent the inequality of wealth distribution. The figure below shows a typical Lorenz curve. The area enclosed between the straight line and the curved line is called the Gini coefficient. The further away the curved line is from the straight line, the more inequality in the wealth is indicated.

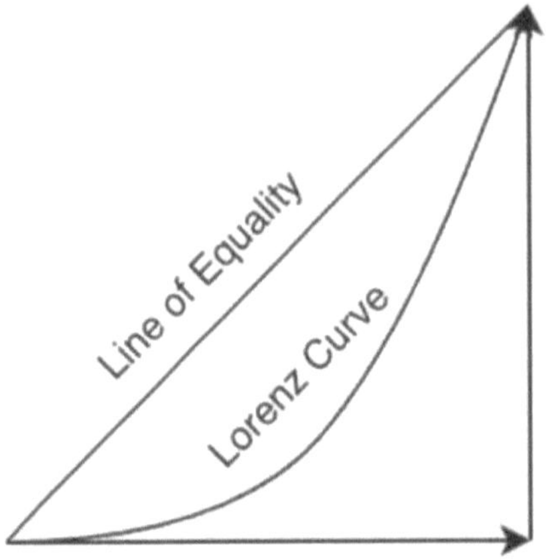

This curve is used in a lot of fields such as ecology, studies of biodiversity, and business modeling.

Note: A graphic method, which estimates the dispersion from shape of a curve, is called Lorenz Curve.

References

1. Burr,L.W., Applied statistical methods, Academic Press, New yark.
2. Cowden, D.J., statistical methods in quality control, prentice Hall, 1957 and Asia Publishing House, 1960.
3. Croxton and Cowden, Applied General statistics, prentice Hall, London and prentice Hall of India.
4. Fisher, R.A., Statistical Methods for research workers, Oliver and Boyd Ltd., London.
5. Gravetter FJ, Wallnau LB. Statistics for the behavioral sciences. 5th ed. Belmont: Wadsworth – Thomson Learning; 2000.
6. Glaser AN. High Yield Biostatistics. 1st Ed. New Delhi, India: Lippincott Williams and Wilkins; 2000
7. Rao PS Sundar, Richard J. Introduction to biostatistics and research methods. 4th ed. New Delhi, India: Prentice Hall of India Pvt Ltd; 2006
8. Marini et al. (2004). Gingival Recession: Prevalence, Extension and Severity in Adults.
9. Swinscow TD, Campbell MJ. Statistics at square one. 10th ed. New Delhi, India: Viva Books Private Limited; 2003
10. Sharma A.K., Text Book of Elementary Statistics, 2005, Discovery Publishing House, New Delhi.
11. Wall, M., Boen, J. & Tweedie, R. (2001). An effective confidence interval for the mean with samples of size one and two. The American Statistician. Web.